IMAGES
of America

Craters of the Moon National Monument

The young lava flows of Craters of the Moon stand out against the older lava terrain that defines the eastern Snake River Plain of Idaho. Craters of the Moon is the largest post–ice age basaltic lava field in the contiguous United States. (Courtesy of NASA.)

On the Cover: Robert Limbert poses next to a "lava tree," formed when flowing lava solidified around a living tree. This image originally appeared in the article "Among the 'Craters of the Moon' " in the March 1924 edition of *National Geographic*. Written and illustrated by Limbert, the article brought national attention to the remote lava fields of Idaho. (Courtesy of the Robert Limbert Papers, Boise State University.)

IMAGES
of America

CRATERS OF THE MOON NATIONAL MONUMENT

Ted E. Stout
Foreword by Jonathan B. Jarvis

ISBN 978-1-4671-0829-4

Published by Arcadia Publishing
Charleston, South Carolina

Printed in the United States of America

Library of Congress Control Number: 2022943162

For all general information, please contact Arcadia Publishing:
Telephone 843-853-2070
Fax 843-853-0044
E-mail sales@arcadiapublishing.com
For customer service and orders:
Toll-Free 1-888-313-2665

Visit us on the Internet at www.arcadiapublishing.com

To all who preserve, protect, and provide for the enjoyment of our national parks.

Contents

FOREWORD

In the fall of 1991, my wife, Paula, and I packed up the family and moved from North Cascades National Park to Craters of the Moon National Monument. When we stopped the car at the first major overlook of the broken and black lava fields, one of our kids exclaimed, with some sense of worry: "Are we going to live here?" We did live there for the next three years, nestled into park housing among the cinder cones, mule deer, big-eared bats, yellow-bellied marmots, limber pines, and starry night skies. The startling scenery of Craters of the Moon often evokes emotional responses of wonder, curiosity, and occasionally revulsion, such as that from Oregon Trail travelers who bumped their wagons along the rocks over 150 years ago. Such wonder stimulated the first National Park Service director, Stephen Mather, to recommend to Pres. Calvin Coolidge in 1924 to use the Antiquities Act (Public Law 59-209) to protect the core of the largest young lava flow in the contiguous United States. The proclamation states that "this area contains many curious and unusual phenomena of great educational value and has a weird and scenic landscape, peculiar to itself." As the new park superintendent, I was struck by how the monument looked like one small jigsaw piece sawn out of a geological puzzle so large it can be easily seen from space. It took until 2000, when Pres. Bill Clinton used the same Antiquities Act to expand that protection to over 750,000 acres to include the entire Great Rift. His proclamation stated, "The most recent eruptions at the Craters of the Moon took place about 2,100 years ago and were likely witnessed by the Shoshone people." Ted Stout, with whom I worked at Mount Rainier National Park, leads you on a visual journey through this place, offering insights into the natural and cultural history, stories from the Shoshone people, adventures of early explorers, and the evolution of the monument's management and protection. It is an invitation to experience this peculiar landscape on your own. We have learned from the astronauts who visited both Craters of the Moon and the real moon that they are different, but walk out on the vast cinder fields, stare across the uninterrupted sweep of the Great Rift, and you can imagine you are in the Sea of Tranquility.

—Jonathan B. Jarvis
Superintendent, Craters of the Moon National Monument, 1991–1993
Eighteenth director, National Park Service, 2009–2017

ACKNOWLEDGMENTS

Without the support of many individuals and organizations, this national monument, as well as this book, would not have been possible. My heartfelt thanks go out to cultural resources program manager Jared Infanger for patiently working with me to provide access to park archives and images. Retired chief of interpretation (1978–2003) Dave Clark provided encouragement and editing. Both he and filmmaker Steve Wursta were also responsible for rescuing important images of Robert Limbert's expeditions that are featured in this work. Other National Park Service (NPS) colleagues, including Lennie Ramacher, Doug Owen, John Apel, Todd Stefanic, and my wife, Rose Rumball-Petre, provided feedback and editing. Tom Blanchard (former Craters of the Moon Natural History Association board member), David Freiberg (Bureau of Land Management), Don Scott (retired NASA educator), and Nolan Brown (historian with the Shoshone-Bannock Tribes Cultural Preservation Department) also provided feedback.

Most of the images originated from the Craters of the Moon archives, but other organizations and individuals also contributed. Photographer Craig Wolfram gave permission to use several of his beautiful shots of the Craters' wilderness. Staff at the Boise State University Library Special Collections and Archives maintain the Robert Limbert Papers and provided many images. Others were obtained from the Idaho State Archives, the Montana State Historical Society, the National Environmental Education Foundation, Northern Arizona University Archives, and the Peabody Museum of Archaeology and Ethnology at Harvard University. The Craters of the Moon Natural History Association provided funding to purchase rights and reproductions for some of these images. Several other images were obtained from the US Geological Survey (USGS), National Aeronautics and Space Administration (NASA), and the Bureau of Land Management (BLM). I greatly appreciate the hard work of the archivists who preserve and make available these important images.

Special thanks are due to David Louter, NPS historian and author of *Administrative History* (1992) and *Historic Context Statements* (1995), which form the foundation for this book. Finally, I am grateful for the excellent foreword provided by former Craters of the Moon superintendent and NPS director Jon Jarvis.

Unless otherwise noted, all images appear courtesy of Craters of the Moon National Monument and Preserve.

INTRODUCTION

[Craters of the Moon is] *an area which contains a remarkable fissure eruption together with its associated volcanic cones, craters, rifts, lava flows, caves, natural bridges . . . which are of unusual scientific value . . . contains many curious and unusual phenomena of great educational value and has a weird and scenic landscape, peculiar to itself.*

—Pres. Calvin Coolidge
Presidential Proclamation No. 1694, May 2, 1924

Remarkable, *unusual*, *curious*, *weird*, *scenic*, and *peculiar* are some of the terms that President Coolidge used when he proclaimed a new national monument in Idaho in 1924. Anyone who has encountered this landscape can certainly identify with these sentiments and may be inspired to add even more colorful adjectives to the mix. The landscape of Craters of the Moon evokes strong reactions.

"This place is sacred to us, very sacred," said Lavern Broncho, Shoshone-Bannock elder, in 2012. "What's unique is that this landscape is protected. This place may be the last stronghold of our native plants."

Sacred is a term that the Shoshone and Bannock people (Sho-Ban) have used to describe this area, part of their ancestral lands. Their ancestors witnessed the eruptions that created the lava landscape and they passed those stories down through the generations. They moved through the area on seasonal migrations, leaving behind evidence of their passage with well-worn trails and ancient rock structures.

The celebrated American writer Washington Irving put forth a very different perspective in his 1837 book *The Adventures of Captain Bonneville*. Bonneville had traveled across the West in 1832, including a passage through the Snake River Plain. Irving wrote: "Nothing meets the eye but a desolate and awful waste; where no grass grows nor water runs, and where nothing is to be seen but lava."

Julius Merrill was one of many Oregon Trail emigrants who recorded negative reactions to the lava landscape while following an alternate route, Goodale's Cutoff, through the area. Merrill was more than happy to leave behind the treacherous terrain of sharp lava rock and continue west. "It was a desolate, dismal scenery," he wrote in 1864. "Up or down the valley as far as the eye could reach . . . the same unvarying mass of black rock. Not a shrub, bird, or insect seemed to live near it. Great must have been the relief of the volcano, powerful the emetic, that poured forth such a mass of black vomit."

Eventually, curiosity about this blank spot on the map led others to purposely seek it out. Geologists, including Harold Stearns and his wife, Norah, described the area in reverent terms. They were excited to view the well-preserved lava features and speculate on their formation and age. Norah Stearns proclaimed in 1928, "The mountains give one a sense of security as belonging to one's own world, yet the lavas black and barren as they are, have a weird charm. The mountains seem to be a symphony, the lavas a rhapsody!"

Unearthly is how Boise taxidermist Robert Limbert described the area: "Right here in Idaho . . . a miniature replica of the moon country . . . a riot of color and fantastic shapes so unearthly as to make one believe himself on another planet," he wrote in the 1920s. Limbert explored and became an ardent promoter of the region. His advocacy, along with scientific justifications provided by Harold Stearns, led to the establishment of Craters of the Moon National Monument as a unit of the National Park System on May 2, 1924.

After reading Limbert's article "Among the 'Craters of the Moon' " in the March 1924 *National Geographic,* the founding director of the National Park Service, Stephen Mather, expressed support for the new monument: "There is not the slightest doubt . . . this area is national monument material of the highest interest." However, the first caretaker, or custodian, Samuel Paisley, was not appointed until a year after the monument's establishment. Paisley was paid $1 per month and had to use his own funds to create a base of operations and improve access for visitors.

It took another year before the monument received its first operating budget of $560. Not until 1935, during the tenure of the fourth custodian, was an additional employee, a seasonal park ranger, added to the staff. Thus began a long-term pattern of neglect for this out-of-the-way monument. With little support, the tiny staff did the best they could to protect park resources and greet increasing numbers of visitors.

Although some funding trickled in from Depression-era programs, substantial support was not provided until the late 1950s through the National Park Service's Mission 66 program. Mission 66 provided critical funding for parks across the nation as they prepared for crowds of post-war travelers. "This is quite a change from the 10' x 14' one room tent covered frame shack which served as office and headquarters for the past twenty-five years," wrote Superintendent Everett Bright in 1958. Besides the construction of the current park headquarters and other badly needed infrastructure, Mission 66 also provided funds for more staff. With new staff onboard, the park began to professionalize the roles of employees and provide more services for the visiting public.

More recent visitors have included Apollo astronauts who toured the area in 1969 to learn about volcanic geology before their lunar voyages. During a subsequent visit to the monument, astronaut Edgar Mitchell stated, "The moon has this peculiar eerie beauty like these flows here . . . that excites your imagination." NASA scientists continue to conduct important research here. Darlene Lim, NASA scientist, wrote in 2019, "this incredible environment is as relevant and scientifically important today as it was fifty years ago. We are here to figure out how best to support human and robotics missions to the moon, and Mars."

In 2000, Craters of the Moon entered a new era when the National Park Service and Bureau of Land Management began to cooperatively manage a greatly expanded monument. The issues today are increasingly complex, but the mission of these agencies to protect and manage the use of nationally significant natural and cultural resources remains the same. On November 9, 2000, Pres. Bill Clinton wrote in Presidential Proclamation No. 7373, "This proclamation enlarges the boundary to assure protection of the entire Great Rift volcanic zone and associated lava features, all objects of scientific interest."

For thousands of years, people have avoided, endured, and pondered this landscape. As we approach the monument's centennial in 2024, consider how your experiences here are part of a legacy of shared responsibility for taking care of this special place.

How do you describe such a place? I invite you to take an armchair tour through this peculiar place by way of this pictorial history, then strap on your boots and experience this weird and scenic landscape on your own terms.

Beginning 15,000 years ago, lava erupted from deep fissures collectively known as the Great Rift. Eight major eruptive periods culminated in the most recent eruption about 2,000 years ago. The Great Rift is hundreds of feet deep and is the deepest known land-based open volcanic rift in the world. At more than 50 miles (80.5 kilometers) in length, it is also one of the longest volcanic rifts in the continental United States. This aerial photograph taken by chief park naturalist Arthur Hathaway in 1965 shows the alignment of the Great Rift through Sheep Trail and Fissure Buttes.

One

Volcanic Wonderland

After several days, thunder and lightning passed over the mountain and aroused the wrath of the serpent. . . . Angered, the serpent began to tighten its coils around the mountain . . . the pressure became so great that the stones began to melt. Fire came from the cracks. Soon liquid rock flowed down the sides of the mountain . . . at last the fire burned itself out, the rocks cooled off, the liquid rock became solid again.

—Shoshone-Bannock oral tradition
Recorded by Ella Clark in *Indian Legends of the Northern Rockies*, 1966

Craters of the Moon National Monument and Preserve contains three separate lava fields formed by volcanic eruptions that originated from a 52-mile-long tear in the earth's crust known as the Great Rift. The Great Rift provided a conduit for magma to reach the surface during eight major eruptive periods beginning 15,000 years ago and continuing until only 2,000 years ago. The resulting volcanic features appear to have happened yesterday and will likely continue tomorrow.

The lava has shaped the ecosystem and cultures of the Snake River Plain. The Shoshone and Bannock people have inhabited the region since at least the end of the last ice age, first showing up in the archeological record 12,000–14,000 years ago. Based on the archeological record and oral traditions, they witnessed some of the eruptions that formed Craters of the Moon.

They passed through the region each year, hunting and gathering plants. Ancient stone structures, well-worn trails, and other artifacts document their extensive travels. Archeological sites include hunting blinds, rock shelters, and rock rings. Artifacts include arrowheads and volcanic glass quarry sites. Some area lava tubes that held ice year-round were used to store meat, like a modern freezer. There are over 5,000 worked stone artifacts, including arrowheads, pieces of pottery, and other artifacts in the park museum collection. Members of the Sho-Ban tribes continue an enduring relationship with the area.

Modern events, such as this 1983 eruption from a volcanic fissure in Hawaii, provide volcanologists with opportunities to learn about past geologic activity at places like Craters of the Moon (Courtesy of USGS.)

As the fissure becomes clogged with hardened lava, the eruption changes from a "curtain of fire" to activity from individual vents. These later eruptions may form cinder cones. A modern example of a cinder cone eruption is shown in Hawaii. (Courtesy of USGS.)

Big Cinder Butte is the tallest cinder cone in the park, rising more than 700 feet (213 meters) above the surrounding landscape. This photograph was taken by former chief park naturalist Edgar Menning in 1964.

There are dozens of cinder cones within the monument and preserve. This is the interior of Echo Crater. Watchman (left) and Sentinel cones are visible in the background. (Courtesy of the Robert Limbert Papers, Boise State University.)

Spatter cones, like this modern example in Hawaii, generally form toward the end of the eruptive cycle when gas-depleted globs of molten lava (spatter) are hurled a short distance from the vent. (Courtesy of USGS.)

Well-preserved examples of these miniature volcanoes are visible along the Loop Road. They were a must-see even before the road or the park were established.

The lava fields contain a remarkable and unusual array of basaltic volcanic features. The dry southern Idaho climate helps to preserve features such as this lava river and ropey pahoehoe flows. The image above was captured by former superintendent Roger Contor in 1965, and the photograph below was taken by Edgar Menning in 1964.

Small-scale features like these oddly shaped lava bombs are widely present. Lava bombs are globs of molten lava that hardened in flight during past eruptions. (Courtesy of the Robert Limbert Papers, Boise State University.)

When flowing lava surrounds a tree, it may harden into the shape of the tree forming a tree mold. Sometimes the tree is knocked down by the flow, forming a horizontal tree mold. If the tree remains standing, it may form a vertical tree mold or lava tree. Most of the time only the imprint of the charred wood remains, but rare charcoal deposits provide an important tool for geologists to carbon-date the flows. (Courtesy of the Robert Limbert Papers, Boise State University.)

Searing lava flows that destroyed everything in their path today surround and protect native plant communities on the Snake River Plain from human activity. Craters of the Moon National Monument and Preserve contains more than 500 of these isolated islands of vegetation that are known by the Hawaiian term *kipuka*. They serve as key benchmarks for the scientific study of long-term ecological change. Round Knoll Kipuka is seen in this 1926 photograph by geologist Harold Stearns. (Courtesy of the USGS.)

One of the kipukas in the preserve shelters enormous juniper trees. The miles of lava that surround this remote grove have allowed these trees to grow unhindered for hundreds of years. (Courtesy of Craig Wolfram.)

Indian Tunnel is named for the rock rings near the cave's entrance that indicate native people visited here. Little is known about the exact purpose of these rings or when they were built; however, they predate the establishment of the monument. These photographs show the interior of Indian Tunnel and the first park custodian amongst the rings in 1926. (Both photographs by Harold Stearns, courtesy of the USGS.)

The lack of water resources prevented the development of permanent villages here. However, ancient stone structures (seen here), artifacts, and well-worn trails across the lava indicate that native people traveled extensively through the region.

Craters of the Moon is part of the ancestral home of the Shoshone and Bannock peoples. From left to right are Charlie Pizoka, Logan Appenay, unidentified, Jack Edmo and one of his sons, and unidentified. This c. 1904 photograph was taken by George Scott at Fort Hall. (Courtesy of the Peabody Museum of Archaeology and Ethnology, Harvard University 2004.29.5348.)

Arrowheads and other projectile points have been surveyed in various locations in the park. Outcrops of glassy lava, known as tachylyte, served as quarries for obtaining raw materials for making some tools.

In 1995, artifacts, including a small amount of burnt bone and rock chips, were unearthed in the campground area. Sho-Ban people spent the night here long before modern campers used this space.

Members of the Shoshone-Bannock tribes continue an enduring relationship with the area. At left is Sho-Ban elder Lavern Broncho during a 2008 visit to the park. Below, students from the Sho-Ban school participate in educational activities at the park. (Left, courtesy of David Clark; below, courtesy of the National Environmental Education Foundation.)

Two

Before the Monument

It is a place of color and silence.

—Robert Limbert, 1923

Most outsiders avoided the lava fields of Idaho; however, in the 1860s, most Oregon Trail emigrants began to follow an alternative route through the area. Goodale's Cutoff was named after Tim Goodale, a mountain man who successfully led 338 wagons from Fort Hall to Boise in 1862. The trail hugged the northern edge of the lava, continuing west along the base of the Pioneer Mountains, paralleling today's US Highway 20/26/93.

Eventually, a few hardy ranchers and miners tried to make a go of it in the northern foothills. An old watering trough and abandoned mine shafts are all that remain today to document these activities within the NPS monument and preserve. The practice of livestock grazing continues today on BLM lands.

Gradually, perspectives shifted when scientists began to explore the area. Geologist Israel C. Russell surveyed the Cinder Buttes area in 1901 and 1902. Harold and Norah Stearns followed up with lengthy visits to the area in 1921, 1923, and 1926. Geologists were impressed by the "freshness" of the lava features and excited to tell others about them. The Stearns continued their research in Hawaii, where they were able to compare the active volcanic processes they observed there to features they had seen in Idaho. Harold wrote the scientific justification for the new monument in 1923 and, soon after, a classic guidebook to the area that was republished for almost 50 years.

Robert Limbert, one of Idaho's most tireless and enthusiastic promoters, began to explore Craters of the Moon in the early 1920s. This blank spot on the map, where stories of lost valleys and grizzly bears abounded, piqued his curiosity. He ventured far into the lava, recording his experiences with his Graflex camera. He shared his adventures in the March 1924 edition of *National Geographic*. His efforts drew national attention to the fascinating volcanic formations here and the need to protect them.

Beginning in the 1850s, an increasing number of Oregon-bound emigrants followed an alternative route, which became known as Goodale's Cutoff. This wagon trail followed a narrow passage between the Pioneer Mountains and the northern edge of the Craters of the Moon lava field. Journal entries indicate that the area left an indelible, but mostly negative, impression on these early visitors. (Courtesy of the Idaho State Archives, P1960-175-6.)

Mrs. W.A. Loughary, a traveler to the area, wrote in her journal in July 1864 that it "could only remind one of the black valley of death. . . . This proved to be far the hardest part of our travel, and yet we must go on or perish by the road side. Every man, woman and child must walk in order to lessen the weight of our axle trees to prevent breaking."

On a ranch not far from the monument, a small mound of rocks sits in the middle of a pasture. Past residents of the ranch say the rocks mark the gravesite of a pioneer girl who never made it to Oregon.

In 1994, park staff hosted a reenactment of a wagon train on the historic trail. Costumed participants portrayed members of the party, including mountain man Tim Goodale and his Shoshone wife, Jennie. In 1862, Goodale guided a group of 1,095 people, 338 wagons, and 2,900 head of stock safely from Fort Hall to Boise. It took this enormous wagon train—the largest to travel any section of the Oregon Trail—over three hours to get into or out of camp.

Most emigrants happily continued westward on the trail, but as time passed, a few hardy folks established ranches and mines in the region. Local rancher Era Martin built this concrete livestock trough near Little Prairie Waterhole around 1920. Livestock grazing continues today outside of the NPS monument and preserve on lands managed by the BLM.

A lead-silver mining boom began in the Pioneer Mountains in the 1880s. The short-lived boomtowns of Era and Martin developed to support these enterprises. Just prior to the establishment of the monument in 1924, the Martin Mine was developed on Little Cottonwood Creek and operated sporadically through the mid-1950s. In 1967, the Craters of the Moon Natural History Association purchased the remaining claim and donated it to the NPS. Structures associated with the mine were removed in the 1980s, with additional landscape rehabilitation completed in the early 1990s.

While completing a reconnaissance of the Snake River Plain for the USGS in the summer of 1901, Israel C. Russell visited a cluster of cinder cones that he referred to as the Cinder Buttes. He became the first geologist to survey and describe the scientific importance of the area. In a subsequent report, he stated that the volcanoes "are remarkably fresh in appearance and furnish most instructive illustrations of the nature of the eruptions which deluged such a large part of southern Idaho with lava." (Photograph by Israel C. Russell, courtesy of the USGS.)

Russell returned the following summer for further research. He described the lava flows as featuring "many pleasing variations in color, ranging from deep red through brown and purple to lusterless black." He named one of these colorful flows Blue Dragon, which he described as appearing like the scales of a reptile. (Photograph by Israel C. Russell, courtesy of the USGS.)

Geologist Harold T. Stearns described the area as appearing like the "surface of the moon as seen through a telescope." He explored and studied the area in the 1920s and became an outspoken advocate for its preservation. This photograph by Stearns shows a part of Two Point Butte. (Courtesy of the USGS.)

At the request of the NPS, Stearns prepared a report in 1923 "describing the area, delineating the boundaries, and stating the reasons for its preservation as a national monument." This photograph by Stearns shows Big Cinder Butte. (Courtesy of the USGS.)

In 1921, Harold and Norah Stearns set up a rough camp on the slopes of Inferno Cone. The Idaho Bureau of Mines and Geology first published their report on the area, along with these photographs, in 1928. This work was republished in various guises through the 1970s. (Courtesy of the USGS.)

Norah D. Stearns accompanied her husband on his research trips to the area. She specialized in hydrology and wrote evocative descriptions of her experiences here. "I always seemed to be gazing out of the door or window in a troubled, puzzled way," she said. "Something was strange! Perhaps this area is indeed an unfinished corner of the universe where the chaos of the primeval world still exists." (Courtesy of the USGS.)

Robert W. Limbert, a taxidermist and photographer from Boise, explored the area extensively in the early 1920s. His enthusiasm for and efforts to publicize the region drew national attention to the fascinating volcanic formations here, as well as the need to protect them. (Courtesy of the Robert Limbert Papers, Boise State University.)

Limbert captured many of his adventures across the lava on his Graflex camera. He also promoted the area through numerous articles that appeared in local and national publications including *Sunset Magazine*, *Outdoor Life*, and a guidebook published for the Union Pacific Railroad entitled *Unknown Places in Idaho.* (Courtesy of the Robert Limbert Papers, Boise State University.)

Limbert's colorful descriptions captured the attention of many. In the *Idaho Statesman* in 1921, he wrote, "Right here in Idaho . . . a miniature replica of the moon country, a vast expanse, silent, dead except for the occasional bird, a country with cold volcanic mountains, numberless 'sputter-cones,' bottomless pits, frozen lakes and rivers and waterfalls of stone, hillsides of flaming red, cliffs of cobalt blue, cones striped with yellow and green, a riot of color and fantastic shapes so unearthly as to make one believe himself on another planet." (Courtesy of the Robert Limbert Papers, Boise State University.)

A Limbert expedition member takes in the view from Echo Crater looking to the south. (Courtesy of the Robert Limbert Papers, Boise State University.)

Robert Limbert originated many of the names that appear on maps of the area today, including Echo Crater (above) and Vermillion Chasm (below). (Both, courtesy of the Robert Limbert Papers, Boise State University.)

In May 1920, Limbert hiked the 52-mile Great Rift from south to north with his friend Walter Cole and his trusty Airedale terrier Teddy. The sharp lava rock took its toll on the feet and paws of the party, and they sought refuge in Echo Crater for a few days to heal. (Courtesy of the Robert Limbert Papers, Boise State University.)

Slowed by the lava terrain and the party's desire to explore and photograph the area, the journey lasted 17 days. At the end of the trek, they were met with a welcoming party near today's monument headquarters area. Music from a local band, a picnic, and speeches rounded out the celebration. (Courtesy of the Robert Limbert Papers, Boise State University.)

In June 1921, Limbert led his most famous investigation of the area. The expedition party consisted of 10 men "equipped to make an exhaustive study of the lava formations, birds and animal life, and explore the many craters." Among the members were two biologists, Luther Goldman of the US Biological Survey and W.E. Crouch of the Smithsonian Institution. The expedition mapped various features, including craters, pits, ice caves, and natural bridges. (Courtesy of the Robert Limbert Papers, Boise State University.)

Other notable participants on the 1921 expedition included Arco residents Samuel Paisley (far right) and Clarence Bottolfsen (top). Paisley would later be appointed the first manager of Craters of the Moon in 1925, and Bottolfsen served two terms as governor of Idaho, in 1939 and in 1943. (Courtesy of the Robert Limbert Papers, Boise State University.)

In later years, Limbert traveled around the country promoting the wonders of Idaho with a varied program of trick shooting, bird calls, slide presentations, and storytelling. He met the expectations of city-dwellers by wearing the garb of a movie cowboy while on tour. Chicago booking agent Dorothy Fox characterized him as an "entertainer deluxe, an extraordinary naturalist, explorer, author, photographer, and cowboy humorist—second only to Will Rogers." (Courtesy of the Robert Limbert Papers, Boise State University.)

The highlight of many of Limbert's entertaining shows was a demonstration of his amazing shooting skills and remarkable marksmanship. He gained the stage name "Two-Gun Bob" because of these skills. While on tour in Chicago, he challenged Al Capone to a duel. The notorious mob boss, impressed by his courage, politely invited Limbert to a dinner engagement, where no guns were reportedly drawn. (Courtesy of the Robert Limbert Papers, Boise State University.)

In March 1924, *National Geographic* published an article entitled "Among the 'Craters of the Moon.' " The article, written and illustrated by Limbert, brought this unknown area to national attention. Limbert also sent a scrapbook of photographs to Pres. Calvin Coolidge to encourage the president to establish a national park in Idaho.

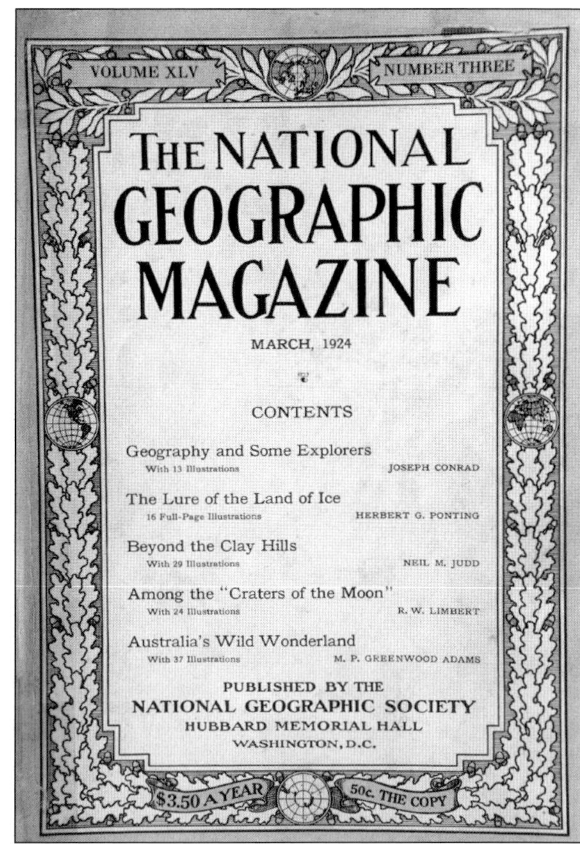

VOLUME XLV NUMBER THREE

THE NATIONAL GEOGRAPHIC MAGAZINE

MARCH, 1924

CONTENTS

Geography and Some Explorers
With 13 Illustrations JOSEPH CONRAD

The Lure of the Land of Ice
16 Full-Page Illustrations HERBERT G. PONTING

Beyond the Clay Hills
With 29 Illustrations NEIL M. JUDD

Among the "Craters of the Moon"
With 24 Illustrations R. W. LIMBERT

Australia's Wild Wonderland
With 37 Illustrations M. P. GREENWOOD ADAMS

PUBLISHED BY THE
NATIONAL GEOGRAPHIC SOCIETY
HUBBARD MEMORIAL HALL
WASHINGTON, D.C.

$3.50 A YEAR 50c. THE COPY

In 1928, Limbert purchased a small hotel and boating operation near Stanley in the Sawtooth Mountains. He expanded this facility, leading to the creation of the venerable Redfish Lodge on the shores of Redfish Lake. (Courtesy of the Robert Limbert Papers, Boise State University.)

In May 1926, Robert Limbert (far right) led his last major expedition in search of the legendary Lost Valley. He was accompanied by a group of hardy adventurers, including five women, recruited from Washington mountaineering clubs. Here, Limbert leads the way followed by J.N.O. Thomson, Clarence Starcher, Herman Krug, Roland Whitmore, Olta Welshone, Joseph Yolo, Athene Clymer, and Elsie Hanft. (Photograph by Joseph Yolo, courtesy of the Robert Limbert Papers, Boise State University.)

The group hiked south for many miles along the Great Rift, ultimately arriving at a lone cinder cone. There they found numerous signs that Sho-Ban people had spent time in this far-flung outpost. They also discovered ice-floored caves that provided water for the group. (Photograph by Joseph Yolo, courtesy of the Robert Limbert Papers, Boise State University.)

Even before the monument was established, curiosity led a variety of folks to explore the area. Thanks to the efforts of Limbert, Harold and Norah Stearns, and others, public appreciation and visitation began to grow. The photograph below was taken in 1918, but the date of the image above is unknown.

Three

Establishment and Formative Years

The water situation is going to be one of the most serious problems to confront us.

—Samuel Paisley
Custodian, Craters of the Moon National Monument, 1926

On May 2, 1924, Pres. Calvin Coolidge established Craters of the Moon National Monument. On June 15, local supporters gathered on site to celebrate the opening of the new monument. The first NPS officials did not visit until September. Finally, on June 1, 1925, Arco resident Samuel Paisley began serving as the first custodian and sole representative of the agency. Thus was established a tradition of neglect for this isolated national monument.

Without an operating budget, Paisley used his own funds to build a small shack near a waterhole that served as the focal point for visitors. By default, this structure became the first headquarters for Craters of the Moon National Monument. Unfortunately, the waterhole near the registration booth dried up two years later.

In hopes of obtaining access to a reliable water supply, new custodian Robert Moore moved headquarters north, to the location of the present campground. In this same area, a local entrepreneur obtained permission to build a lodge and cabins. This development, known as the Crater Inn, was never a grand lodge but it did provide essential services during these formative years. The most essential service was water, which was trucked in and provided to thirsty staff and visitors. Finally, in 1931, water was piped in from springs located on newly acquired land in the Pioneer Mountain foothills. In the early 1930s, the NPS also used Depression-era emergency relief funds to build a house, public restroom, and maintenance shed in this area.

Monument custodians served alone until 1935, when a seasonal employee, ranger G. Frederick Shepherd, was hired. The first year-round park ranger, Robert Zink, was not hired until 1952. With new staff and minimal funding, park officials were forced to construct a rag-tag collection of shacks to provide office space and housing. Many of these structures were kept in service into the late 1950s.

CRATERS OF THE MOON NATIONAL MONUMENT

IDAHO

By the President of the United States of America

A Proclamation

WHEREAS, there is located in townships one south, one and two north, ranges twenty-four and twenty-five east of the Boise Meridian, in Butte and Blaine Counties, Idaho, an area which contains a remarkable fissure eruption together with its associated volcanic cones, craters, rifts, lava flows, caves, natural bridges, and other phenomena characteristic of volcanic action which are of unusual scientific value and general interest; and

WHEREAS, this area contains many curious and unusual phenomena of great educational value and has a weird and scenic landscape peculiar to itself; and

WHEREAS, it appears that the public interest would be promoted by reserving these volcanic features as a National Monument, together with as much land as may be needed for the protection thereof.

NOW, THEREFORE, I, Calvin Coolidge, President of the United States of America, by authority of the power in me vested by section two of the act of Congress entitled, "An Act for the preservation of American antiquities," approved June eighth, nineteen hundred and six (34 Stat., 225) do proclaim that there is hereby reserved from all forms of appropriation under the public land laws, subject to all valid existing claims, and set apart as a National Monument all that piece or parcel of land in the Counties of Butte and Blaine, State of Idaho, shown as the Craters of the Moon National Monument upon the diagram hereto annexed and made a part hereof.

Warning is hereby expressly given to all unauthorized persons not to appropriate, injure, destroy or remove any feature of this Monument and not to locate or settle upon any of the lands thereof.

The Director of the National Park Service, under the direction of the Secretary of the Interior, shall have the supervision, management, and control of this Monument as provided in the act of Congress entitled, "An Act to establish a National Park Service and for other purposes," approved August twenty-fifth, nineteen hundred and sixteen (39 Stat., 535) and Acts additional thereto or amendatory thereof.

In Witness Whereof, I have hereunto set my hand and caused the seal of the United States to be affixed.

[SEAL.] DONE in the City of Washington this 2d day of May in the year of our Lord one thousand nine hundred and twenty-four and of the Independence of the United States of America the one hundred and forty-eighth.

CALVIN COOLIDGE

By the President:
CHARLES E. HUGHES
Secretary of State.

[No. 1694.]

On May 2, 1924, Pres. Calvin Coolidge established Craters of the Moon as a new national monument.

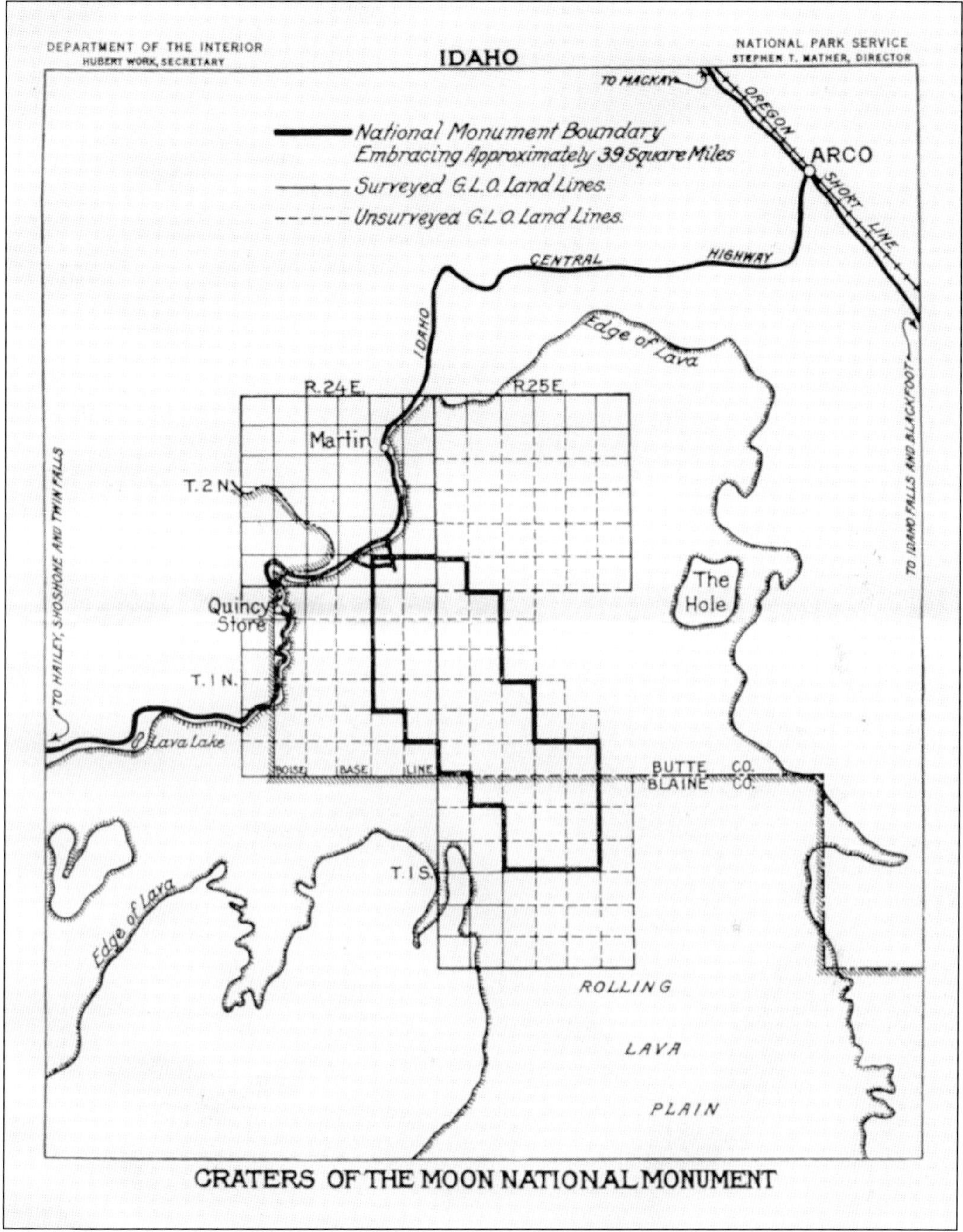

Utilizing the authority of the Antiquities Act, Coolidge designated 39 square miles (101 square km) of federal land in southern Idaho for protection by the National Park Service.

To celebrate the opening of the monument, approximately 1,500 people representing the many individuals and groups who supported the area's designation gathered on the southeast slope of Inferno Cone on June 15. The *Arco Advertiser* reported that "it puts the community on the map

as one of the scenic districts in the west." (Photograph donated by the Paisley-Walker family, courtesy of NPS.)

Four months after the monument's creation, Horace Albright (center, second row), superintendent of Yellowstone National Park and field assistant to the NPS director, conducted the first official inspection of the new site. He was met by local and state officials, most notably local guide Samuel A. Paisley (first row, far left). For the first year of the new monument's existence, there was little tangible support provided by the federal government. Finally, in 1925, Paisley was appointed the first custodian, or manager, of the area. (Courtesy of the Haynes Collection, Montana State Historical Society.)

Paisley cared about the volcanic area and explored and guided others through it for 14 years. He accompanied both Harold Stearns and Robert Limbert during their visits to the area. This image shows Paisley, second from left, providing support for another expedition member in Vermillion Chasm. Initially rejecting the offer of $12 per year with no operating budget, Paisley requested permission to supplement his income by continuing to operate a guide service. NPS director Stephen Mather bent the rules and agreed to Paisley's request in order to place the Park Service's "official representative at the monument." (Courtesy of the Robert Limbert Papers, Boise State University.)

Without an operating budget, Paisley used his own funds to build this structure on the saddle between North Crater and Paisley Cone. It served as a combination on-site residence, contact station, and first monument headquarters.

Arriving visitors were asked to register at this station, which was conveniently located near a waterhole. The bucket in the man's hand was likely used to scoop up some water. Registration Waterhole was one of the few reliable places to obtain water in the park until 1927.

Nearby Cinderhurst campground was the first formal campground in the monument. A flat spot on the cinder field, an outhouse, and access to Registration Waterhole were the only amenities.

The availability of water in this lava desert has always been an issue. In rare instances, deep cracks filled with ice retain water near the surface, preventing it from being sucked down through the porous sponge-like lava rock. During the early years, both visitors and staff relied upon scarce waterholes like Big Sink Waterhole, shown here, for refreshment.

Maintaining sanitary conditions was also problematic. Staff posted signs asking visitors not to wash here. The reliability of these communal water sources also became an issue. In 1927, Registration Waterhole suddenly went dry.

In the early days of the monument, just getting to the park was a test of endurance. Upon arrival, navigating the loose cinder road was another challenge. Custodian Paisley is shown here, on the right in his trademark hat, helping to extricate a wayward motorist. Geologist Norah Stearns described a similar situation she experienced: "When driving into this spell-bound fragment of the universe, let the driver keep his eyes on the road. Too much scenery brought me woe."

When Registration Waterhole dried up, park officials moved headquarters closer to the highway. A variety of structures were eventually built here, in the location of the current park campground, to provide visitor services and housing for staff. The custodian's residence (left) and the Crater Inn (right) were the most prominent of these.

"Comforts come to Craters of the Moon. . . . Cabins and a lunch service free the visit to the lava formations from inconvenience and hardship," announced proprietor Jo G. Martin in 1927. The rustic Crater Inn provided basic services for park visitors including lodging, food, and a gas station. The log walls and lava rock chimney both leaked, allowing cinders and smoke to permeate the lodge. For the first four years of business, water trucked in by the proprietor provided the sole source of drinking water for visitors and staff.

In the 1930s, New Deal funding provided the first substantial development of administrative and visitor facilities in the park. These structures included the log comfort station (above), a log equipment shed, and the custodian quarters (below). The comfort station and the equipment shed are used to this day for their original intended purposes. The log cabin served as the residence for the custodian/superintendent until it was removed in the late 1950s.

Completed in 1932, the log equipment shed is the oldest existing structure in the park. A sign indicating that this was a public works project still graces the interior of the building (below).

Shortly after starting his tenure as custodian in 1927, Robert B. Moore initiated the transfer of park headquarters to the current campground area. He also commenced the development of the park's first water system, which would carry water from springs in the foothills of the Pioneer Mountains, starting in 1931.

In 1931, custodian Burton C. Lacombe transferred from Yellowstone, where he had served as the buffalo keeper. He was the first career NPS employee at the park. Upon his arrival, he and his wife were obliged to shelter in a tent until the new custodian's cabin was completed later that summer.

From 1933 to 1936, custodian Albert T. Bicknell oversaw the completion of the new comfort station, improved the Loop Road, and built the first formal trail to the caves area using New Deal funding.

In 1935, Bicknell hired a seasonal park ranger named G. Frederick Shepherd (likely on the right). Shepherd was a graduate student in geology from the University of Chicago. He researched the volcanic terrain throughout the summer and at the end of the season, he hiked solo across the Great Rift. On his return north, this enterprising young geologist hitched a ride on an airplane.

After New Deal funding ran out, desperation led park managers to construct a variety of ramshackle structures to serve as office and quarters for staff. The structure at right was the main park office and served as the contact station for visitors in the 1950s. Rangers were offered housing in the tarpaper shack nearby (below). These temporary structures were finally removed in 1957 when headquarters was moved a third time.

Since the monument was isolated by snow much of the year, a special Opening Day celebration was held each spring from the late 1920s until the 1970s. These popular community celebrations included picnics, speeches, music, and much fanfare. These photographs show the popularity of the event in 1939 (above) and a decade later (below).

This early postcard highlights some of the scenic natural wonders of the monument.

The Crater Inn and cabins (left) and the custodian quarters are visible in this early postcard view. These structures reflected the rustic style that typified national park construction during this era, but not everyone was happy with the new look. "Log structures are entirely inappropriate here as there is hardly a tree in the whole area," exclaimed custodian Guy McCarty around 1937.

Most of the Loop Road remained an unpaved cinder route into the mid-1950s. Wind eroded the cinder surface, causing various road hazards and a need for constant replenishment of the road base.

This early checking station stood close to the location of the current entrance station in the campground area. The sign mentions "cabins available at the Craters Inn," which are visible to the east of the station.

Robert Zink served as the first permanent park ranger, from 1952 to 1956. As a generalist ranger, he divided his time between providing resource protection and naturalist programs. He also researched and wrote the first history of the monument. Zink's 1954 *Short History* and the 1992 *Administrative History* by NPS historian David Louter provided background for this book.

Lack of funding and resources led to difficult times for park management during and after World War II. By the late 1950s, the monument's infrastructure consisted of a collection of log buildings constructed in the 1930s and a variety of temporary shacks of more recent vintage. As the country emerged from the war years, a flood of visitors began to seek out their national parks.

Four

Mission 66

Mission 66 has broken over our heads, almost with a vengeance, we are among the first areas to receive its many blessings.

—Robert Zink
Ranger, Craters of the Moon National Monument, 1956

Years of low funding combined with a postwar boom in visitation and the upcoming 50th anniversary of the National Park Service gave birth to the 10-year Mission 66 project. Across the country, the NPS built new infrastructure, including roads, campgrounds, and visitor centers. With facilities that were little changed since the 1930s, few parks needed this support more than Craters of the Moon.

By the late 1950s, the monument's structures consisted of various log buildings, wood-frame cabins, converted tent cabins, and a ramshackle collection of outbuildings. The cinder-covered Loop Road remained unpaved, and utility systems were strained to their limit.

A concurrent rise in visitation compounded the poor state of the park's facilities. During the war years, annual visitation hovered around 2,000. By 1954, it had risen to nearly 100,000, and by 1966, it doubled again to nearly 200,000. Americans were anxious to get out and rediscover their national parks.

In preparation for the construction of a new administrative site, several of the old structures were dismantled or removed. The Crater Inn's buildings were sold at public auction, and the custodian's cabin was torn down along with various "temporary" structures that had served the park for many years. Only the sturdy log comfort station and equipment shed survive today as reminders of the monument's formative years.

Paving of the Loop Road commenced in 1956. Construction of the monument's third, and current, administrative site began in 1957 and was completed the following year. The site included a new visitor center/headquarters building, a new maintenance facility, and housing for park staff. A new wastewater system and upgrades to the campground were also completed soon after.

In addition to new workspaces, Mission 66 also provided funding for new staff. Maintenance, administrative, and interpretive staff were hired, leading to the professionalization of the monument's workforce.

Beginning in the late 1950s, Congress appropriated much-needed funds for the National Park Service's Mission 66 program. With these funds, the NPS built new infrastructure across the country.

Mission 66 transformed facilities at Craters of the Moon from a random collection of worn-out structures into a modern, comprehensively planned park. Before construction began, several of the older structures were dismantled or removed from the park including the custodian quarters and the Crater Inn. The Crater Inn main building was sold at auction and relocated to a ranch in the Lost River Valley.

This 1957 image shows the incomplete utility building (center) and staff housing (right). The new visitor center/park headquarters building had not yet been completed, but the foundation and parking lots for the new structure are visible on the left.

In this close-up image, the new utility building begins to take shape. It would provide six garage stalls, a supply room, a shop, and a restroom for park maintenance staff.

Craters of the Moon's new visitor center/park headquarters was completed in 1958. It is an excellent example of the Park Service Modern architectural aesthetic. The view of the compact developed area, including employee housing, has changed little since construction was completed (below).

Even before the establishment of the monument in 1924, visitors had roughed out the basic route of the Loop Road. However, the road remained a rough cinder track until it was paved for the first time in 1957 as a part of the Mission 66 initiative.

Crews pave the Loop Road on the south side of Inferno Cone. Crews also built and paved nine parking areas, installed water and sewage systems, and improved the campground by leveling sites and paving the campground loops.

In June 1958, a ceremony was held to dedicate the new visitor center/park headquarters building. Idaho governor C.A. Bottolfsen (speaking), Supt. Everett Bright (far left), and Secretary of the Interior Fred Seaton (first row, standing far right) were all in attendance.

In 1990, the visitor center was rededicated to park founder Robert W. Limbert to recognize the pivotal role he played in exploring, promoting, and helping to establish the national monument. Although Limbert died unexpectedly in 1933 at the age of 48, his legacy endures in this place he loved.

This view of the visitor center, entrance station, and campground was featured in an article entitled *Man on the Moon* in the October 1960 issue of *National Geographic*. (Photograph by Bill Belknap, courtesy of Cline Library, Northern Arizona University.)

In the 1950s, the "visitor center" was an entirely new building type conceived by NPS planners to welcome the postwar visitation boom. It concentrated visitor services into one structure that served as the information center and included restrooms, museum exhibits, audio-visual presentations, and administrative offices. The NPS built more than 100 visitor centers during the Mission 66 era, including the current facility at Craters of the Moon. (Photograph by Bill Belknap, courtesy of Cline Library, Northern Arizona University.)

Everett Bright witnessed tremendous changes during his tenure as superintendent from 1953 to 1958. He helped plan and guide the park through the initial phases of Mission 66.

Floyd Henderson led the park into a new era during his tenure as superintendent from 1958–1961. He expanded the workforce and laid the groundwork for the 1962 Carey Kipuka expansion.

Along with new workspaces, Mission 66 also brought new staffing to the park. For the first time in the park's history, essential employees, such as maintenance workers, joined the workforce. Maintenance employee James Sipe (pictured) began his tenure in 1953 and worked at the park through 1959. The first maintenance foreman, Floyd Standlee, was hired in 1966.

Hugh Higginbotham was hired as the first administrative clerk in 1957. In 1965, an administrative officer, Mildred Standlee (pictured), was brought on to handle increasingly complicated budgeting.

Superintendent Henderson also hired the first year-round park naturalist, David Ochsner, in 1959. He enhanced many of the practices of earlier, part-time rangers, including the regular presentation of evening programs and guided walks. He also helped to establish the park's cooperating association, the Craters of the Moon Natural History Association, in 1959. This nonprofit association continues to provide critical support for educational, interpretive, scientific, and cultural activities offered to the public.

The first seasonal park ranger, G. Frederick Shepherd, was hired in 1935. Other temporary staff were hired sporadically until funding dried up during the war years. Mission 66 funding reinitiated the practice of hiring summer staff. Here is the 1958 park ranger crew (from left to right) Peter Sanchez, Rodney Romney, and Ronald Erickson in front of the new entrance station.

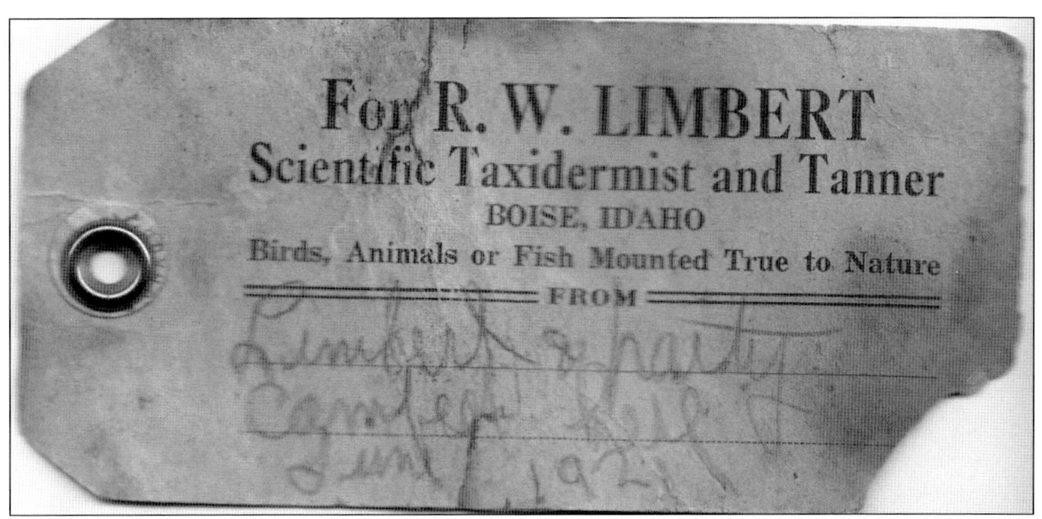

A *National Geographic* photographer, accompanied by Ranger Sanchez, discovered this tag when they bedded down at one of Robert Limbert's former campsites. The items were featured in a photograph in the October 1960 issue of *National Geographic* with a caption reading, "Finder Please Write." This tag is part of the Craters of the Moon museum collection.

The 1961 Opening Day ceremony included an annual event known as the Sheriff's Posse Dash. Flag-bearing horseback riders careened down the ridge east of the campground in this dramatic opening ceremony. Supt. Daniel Davis discontinued this popular spectacle in 1963 due to increasing concerns about resource damage.

Supt. Merle E. Stitt welcomes the Altherr family as the "billionth" visitors to Craters of the Moon on August 22, 1962. The Butte County Chamber of Commerce joined the park for this special recognition, which included prizes for this unsuspecting family from Minnesota. Although the figure was certainly exaggerated, the excitement of staff to welcome new visitors to their shiny new facilities could not be overstated.

Five

Science, Education, and Art

As an individual, I was awed by the serenity of this stark and peaceful landscape. As a geologist, I saw some of the best eruptive fissures in the continental U.S.—a remarkable example of the Earth's energy and fury. The volcanic history of this place suggests another eruption in the near future. When this occurs, the park will instantly attract volcanologists from around the world and become a destination for thousands of curious onlookers.

—Mel Kuntz
Geologist, 2009

As attitudes evolved, scientists, educators, artists, and others began to view Craters of the Moon more as a place to ponder than a place to avoid. The region has long been important for geologic research. Many other scientists followed, investigating other aspects of the natural and cultural history of the area. Botanists have long been fascinated by the isolated kipukas, where relatively pristine islands of sagebrush steppe flora flourish. In 1966, Dr. Earl Swanson led the first systematic survey of archeological sites at the monument. The earliest major wildlife study, on deer populations, was completed in 1983. Since 1969, NASA astronauts and scientists have utilized the monument as an analog for the moon and Mars.

Early management efforts focused primarily on improving roads and trails to improve access and decrease visitor impacts on fragile lava formations. Learning from past missteps and incorporating ecological principles has enabled staff to more proactively monitor and manage park resources in the modern era. Maintaining native plant communities despite increasing wildfires and encroaching weeds is an unceasing task. Important wildlife research continues, with recent studies on pronghorn, bat, and pika populations. Enhanced interpretive and educational efforts have also been critical for providing opportunities for visitors and students to care about and for the park.

Artists of all stripes have long held a fascination for the area. In the early 19th century, authors such as Washington Irving and Mary Hallock Foote brought attention to the lava beds of Idaho. In the 20th century, Robert Limbert and famed landscape photographers Ansel Adams and Philip Hyde focused their lenses on this landscape. In recent years, an artist-in-residence program and partnerships with regional art museums are highlighting new perspectives.

Built by local ranchers, this large bear trap sat near the entrance to the monument in the early years. Up until the 1930s, there were scattered accounts of grizzly bear encounters within the lava fields. In addition, bones discovered in lava pits and tubes provide evidence of their past presence here. Grizzly bears, bighorn sheep, and bison are three species that were historically present but are no longer found on the Snake River Plain.

Robert Limbert was originally drawn to the area by rumors of "dwarf grizzlies" that were said to inhabit the area. Although he never found such a beast, he claimed to have encountered a full-grown grizzly during his 1921 expedition. He captured the "good, old, accommodating bear" with his camera rather than with his rifle and shared the experience in a subsequent article in the *Idaho Statesman*. (Courtesy of the Robert Limbert Papers, Boise State University.)

In the early days of NPS management, it was not uncommon for government hunters to target predatory species. Surprisingly, early monument officials may have also targeted the lowly porcupine with control efforts, in order to prevent their possible impact on park trees. Recently, porcupines have again been sighted within the monument, but they were never very common in this sparsely forested place. (Courtesy of the Robert Limbert Papers, Boise State University.)

The geology of Craters of the Moon has created unique and unexpected habitats that provide for the survival of a surprising diversity of wildlife. Scientists have observed close to 300 different reptiles, birds, and mammal species within the monument and preserve. Among these is a large herd of pronghorn that annually migrate across eight miles of lava through a pinch point between Sunset Cinder Cone and the Pioneer Mountains. To facilitate their movement along this ancient migration route, park staff have removed or modified old boundary fencing. (Courtesy of the Robert Limbert Papers, Boise State University.)

University of Idaho ecologists E.W. Tisdale and M. Hironaka first visited a remote area known as Carey Kipuka in 1956. Afterward, they and another colleague, M.A. Fosberg, promoted the protection of this area because of its scientific importance. Surrounded by miles of rough lava, this vegetated island, or kipuka, retained relatively pristine vegetation largely unaffected by grazing. NPS biologist Adolph Murie confirmed their findings and recommended that the area be added to the monument in 1958.

Between 1961 and 1967, park managers engaged in a controversial program to eradicate a tree parasite known as dwarf mistletoe. The removal of this naturally occurring plant was ultimately unsuccessful, but thousands of its host limber pine trees were killed or damaged in the process. Here, a crew burns a slash pile of pruned mistletoe-affected branches. Mistletoe eradication is one example of a small number of cases of direct resource manipulation at the monument. This controversial practice came at a time when the NPS was just beginning to incorporate ecological principles into its management actions.

Brad Griffith, a scientist from the University of Idaho, and NPS staff completed an extensive study of the monument's mule deer herd in 1983. Numerous deer were tagged and radio-collared in order to track the herd's movements (below). The deer were observed to have a unique dual migration pattern following seasonal moisture and vegetation from the Pioneer Mountains into the desert and back again.

One of the recommendations of Griffith's deer study was to replace the manicured green lawns in the developed area with native vegetation. The green grass and sprinklers were luring deer across the busy highway, resulting in numerous road kills (below).

In the early 1990s, park staff began the process of replacing the green lawns with native plants. The transition back to native plants provided multiple benefits, including fewer deer mortalities on the highway and improved use of the park's limited water supply.

Native plants now surround the visitor center. Interpretive signs on a short nature trail introduce some of the most common plants of the sagebrush steppe ecosystem to visitors before they venture further into the park.

Nonnative plants replace native plant communities, degrade wildlife habitats, and reduce the biological diversity of ecosystems. Non-native plants, including state-listed noxious weeds such as rush skeletonweed and leafy spurge, are managed to prevent them from dominating the landscape. Park staff began limited weed control activities as early as 1968.

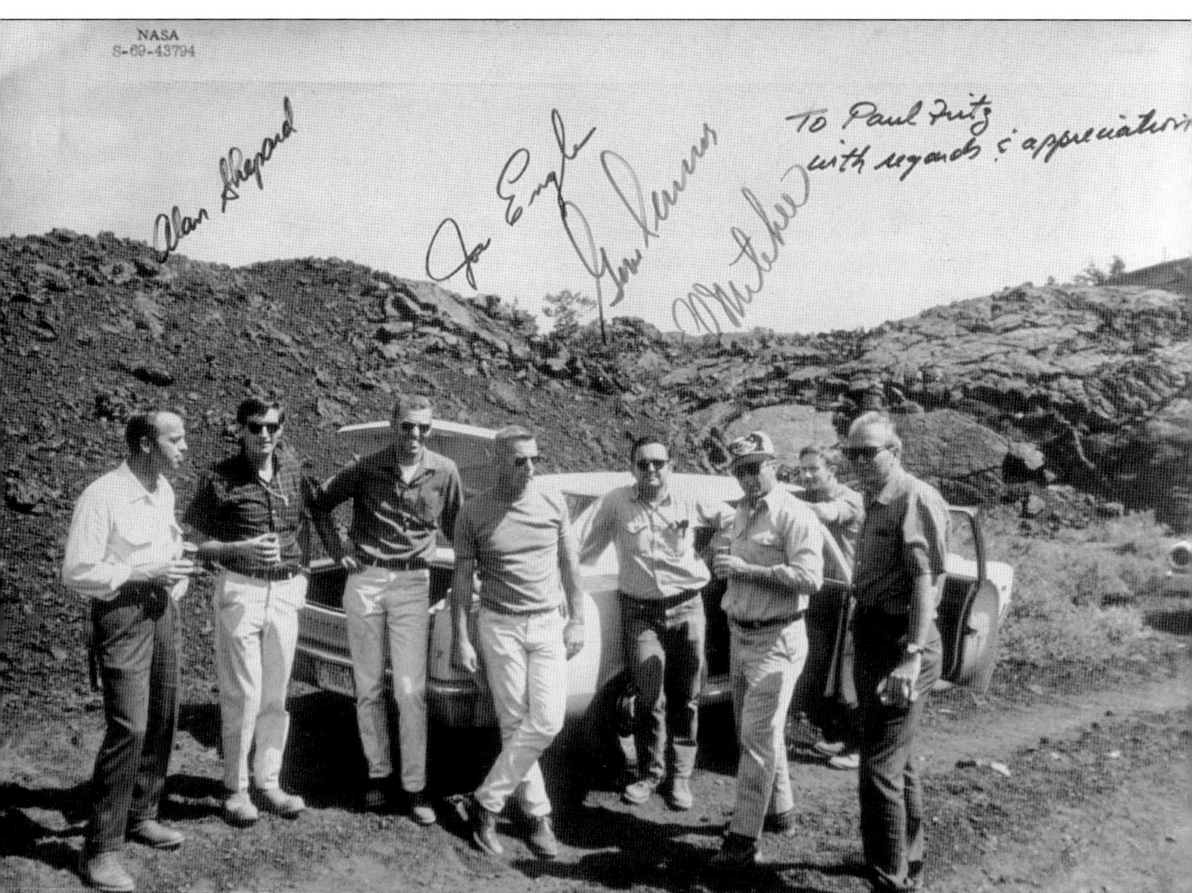

On August 22, 1969, Apollo astronauts Alan Shepard, Joe Engle, Eugene Cernan, and Edgar Mitchell visited Craters of the Moon in preparation for their 1971 lunar mission (Apollo 14). They came here to learn the basics of volcanic geology. Because only a small number of rock samples could be carried home from the moon, NASA officials felt it was important that the astronauts could identify the most scientifically valuable specimens in the field. (Courtesy of NASA.)

Led by NASA geologists and park staff, the astronauts toured the monument and visited other geologically significant areas. Astronauts Edgar Mitchell (left) and Alan Shepard (center) learn about volcanic geology from NASA geologist Mike McEwen. Mitchell served as lunar module pilot for Apollo 14. Shepard was the first American in space on May 5, 1961. He also completed a 15-minute suborbital flight during the Mercury program. Both walked on the moon during the Apollo 14 mission. (Courtesy of NASA.)

From left to right, astronauts Joe Engle, Eugene Cernan, Superintendent Paul Fritz, and geologist Ted Foss explore Buffalo Cave. Engle went on to pioneer the space shuttle program and served as commander of the second and twentieth space shuttle flights. Cernan participated in three different missions, including Apollo 17, where he famously became the last astronaut to walk on the moon. (Courtesy of NASA.)

Apollo 14 commander Alan Shepard is pictured on the lunar surface in February 1971. This was the third Apollo mission to land on the moon. Shepard and Mitchell collected 94.35 pounds (42.8kg) of moon rocks and deployed several scientific experiments. (Courtesy of NASA.)

In 1999, astronauts Mitchell (above), Cernan (below left), and Engle (below right) returned to Craters of the Moon to help celebrate the monument's 75th anniversary, 30 years after they trained here. During their visit and at public presentations, they spoke about how Craters of the Moon helped them to prepare for their missions to the moon.

NASA scientists have visited Craters of the Moon many times to conduct space science research. Beginning in 2015, two large research teams investigated the use of various instruments in the field including Light Detection and Ranging (LIDAR) visualizations of lava tube caves (right) and a model of a next-generation Mars rover (below). (Both, courtesy of NASA.)

In 2016, NASA brought a mission control center (above) to Arco to provide a venue for communicating with simulated astronauts in the field (left). The basalt-covered terrain of Craters of the Moon provides an excellent analog for similar landscapes on Mars. Scientists experimented with communication procedures and various instruments to obtain geologic samples in the field, providing a dress rehearsal for future Mars missions. (Both, courtesy of NASA.)

In 1988, Shelly Sparhawk was appointed the first full-time resource management specialist in the park. In 1992, resource management became its own division led by Vicki Snitzler-Neek (pictured).

The Craters of the Moon Wilderness Area is considered a Class I airshed under the Clean Air Act, which requires that the area receives the nation's highest level of air quality protection. Craters of the Moon began partnering with other federal and state agencies as a part of a national monitoring network in the 1990s. Far from cities, Craters of the Moon retains relatively clean air, which creates excellent opportunities for visitors to view the immense lava landscape and fabulous night skies.

USGS geologist Mel Kuntz mapped and dated the lava flows in the 1980s. He determined that there were eight major eruptive periods over the last 15,000 years. The most recent eruption, about 2,000 years ago (plus or minus 100 years), was determined by carbon dating of charcoal deposits left over from the interaction of lava and vegetation. Our modern understanding of the geologic history of Craters of the Moon is largely based on his research. (Courtesy of David Clark.)

Through a partnership with the Idaho National Laboratory and the USGS, earthquake activity is monitored at this seismic station on the north end of Craters of the Moon. Earthquake activity on the Snake River Plain is relatively rare compared to the surrounding mountains. However, swarms of earthquakes occasionally indicate that magma is stirring beneath the surface. Monitoring this activity may provide an early warning before the next eruption.

Resource management staff have worked with scientists and a network of other NPS units to develop comprehensive inventories of plant and animal species. Long-term monitoring programs are focused on plants (sagebrush steppe, aspen, limber pine), wildlife (bats, pika), water quality, and the impacts of climate change. By monitoring these vital signs over time, the NPS inventory and monitoring program has become a powerful tool for observing and tracking long-term changes affecting Craters of the Moon and other national parks.

Sustainability has become a key consideration for all management decisions at the park. Through improved conservation practices, staff has greatly reduced water and energy use. This photovoltaic array was installed in 2010 and now produces 30–40 percent of the park's energy requirements. (Courtesy of Doug Owen.)

Noting that "this area contains many curious and unusual phenomena of great educational value," President Coolidge highlighted the importance of education in the monument's enabling proclamation in 1924. From the earliest days, park custodians served as guides. Regularly scheduled guided walks began in 1959 when the first full-time park naturalist, Dave Ochsner, was hired. This photograph shows a ranger leading a group on the North Crater Flow Trail around 1963.

Ranger-led auto caravans with stops at various locations were a brief but popular form of interpretation in the early 1960s. Today, most visitors follow the Loop Road on their own, stopping along the way to hike the many trails that lead to important features in the park. Interpretive signs on trails and pullouts provide information for self-guided visits.

Seasonal ranger G. Frederick Shepherd presented the first campfire programs in 1935. These presentations were discontinued during the war years and did not commence again until the 1960s. The photograph above shows a ranger providing an evening program in 1965. Rangers continue to present a wide variety of evening programs in the campground amphitheater (below), which was refurbished in 1988. (Below, courtesy of Jake Frank.)

Park staff started providing on- and off-site presentations to educational groups in the early 1960s. The number of students and schools has continued to increase with each passing year. A wide variety of curricula and workshops have also been offered to "teach the teachers" as well.

Since the early 2000s, park staff has offered guided educational snowshoe walks focused on winter ecology. These popular programs have provided an opportunity for school groups to visit the park beyond the busy fall and spring field trip seasons. (Courtesy of David Clark.)

For more than 25 years, staff partnered with the Idaho Falls Astronomical Society (IFAS) to provide opportunities for viewing the dark skies of Craters of the Moon at annual Star Parties. Today, interpretive staff also provide opportunities for visitors to view and appreciate this valuable resource. Lowell Frauenholz (left), former president of the IFAS, points out a feature in the night sky to visitors during one of these events. (Courtesy of David Clark.)

Devils Orchard, North Crater Flow, and the Snow Cone trail (pictured) are fully accessible for people with mobility impairments. Park managers have discovered they can simultaneously improve the experience for visitors of all abilities and enhance the protection of fragile lava features along these busy trails.

Landscape artist Cindy Tower was the first artist-in-residence at Craters of the Moon in 2010. Since then, park staff has annually hosted artists from a wide variety of disciplines. During their residency, artists develop works that express varying perspectives while honing their craft in this one-of-a-kind location. The artists also share their talent and works with the public through lectures, workshops, and exhibitions.

To celebrate the National Park Service centennial in 2016, park staff partnered with the Sun Valley Museum of Art to host exhibitions on-site and at the museum in Ketchum. Two large works were commissioned for temporary display at the park, including this lava-tube sculpture created by Seattle artist John Grade. (Courtesy of the Sun Valley Museum of Art.)

Six

Designations and Expansions

This Proclamation enlarges the boundary to assure protection of the entire Great Rift volcanic zone and associated lava features, all objects of scientific interest.

—Pres. Bill Clinton
Presidential Proclamation No. 1694, November 9, 2000

Thirty-nine square miles of Craters of the Moon became a national monument on May 2, 1924, when President Coolidge utilized the Antiquities Act to proclaim it so. With no reliable water supply and a very limited land base, Coolidge attempted to remedy the situation four years later by doubling the size of the monument. Ultimately, access to springs from the foothills north of the highway and a path to pipe the precious water to the developed area were achieved after a series of additional land acquisitions and exchanges. The addition of the 180-acre (73-hectare) Carey Kipuka and intervening lands by Pres. John Kennedy to the southwest of the monument filled in the "original" 53,420-acre (21,618-hectare) Craters of the Moon National Monument in 1962.

In 1970, Pres. Richard Nixon signed a bill that designated the 43,243-acre (17,500-hectare) Craters of the Moon National Wilderness Area within the existing monument boundary. This was the first area in any national park designated as wilderness under the 1964 Wilderness Act.

In 2000, the biggest change to the monument to date occurred when President Clinton expanded its size 13-fold to approximately 753,000 acres (304,728 hectares). The new monument would be cooperatively managed with the BLM, with each agency having authority over separate portions. Two years later, congress redesignated the 410,000-acre (165,921-hectare) NPS expansion area as a national preserve, which allowed hunting to continue on these former BLM lands.

Craters of the Moon's natural and scenic qualities have also been recognized with a variety of other special designations. The Great Rift System was designated a national natural landmark in 1968. In 2012, the 140-mile-long Peaks to Craters Scenic Byway was designated by the State of Idaho on US Highway 93 from Challis to Timmerman Junction. In 2017, the International Dark Sky Association recognized Craters of the Moon as an international dark sky park.

Both Goodale's Cutoff (1974) and the Mission 66 Historic District (2022) have been listed in the National Register of Historic Places. These listings broadened recognition of the significance of the monument's historic resources.

There may be changes yet to come. For 100 years, Idahoans have discussed renaming the national monument as a national park. Although the change would not affect the area's management, it would prompt many more visitors to experience this weird and scenic landscape. For now, Craters of the Moon awaits congressional action to properly acknowledge the nationally significant natural and cultural resources of this peculiar place.

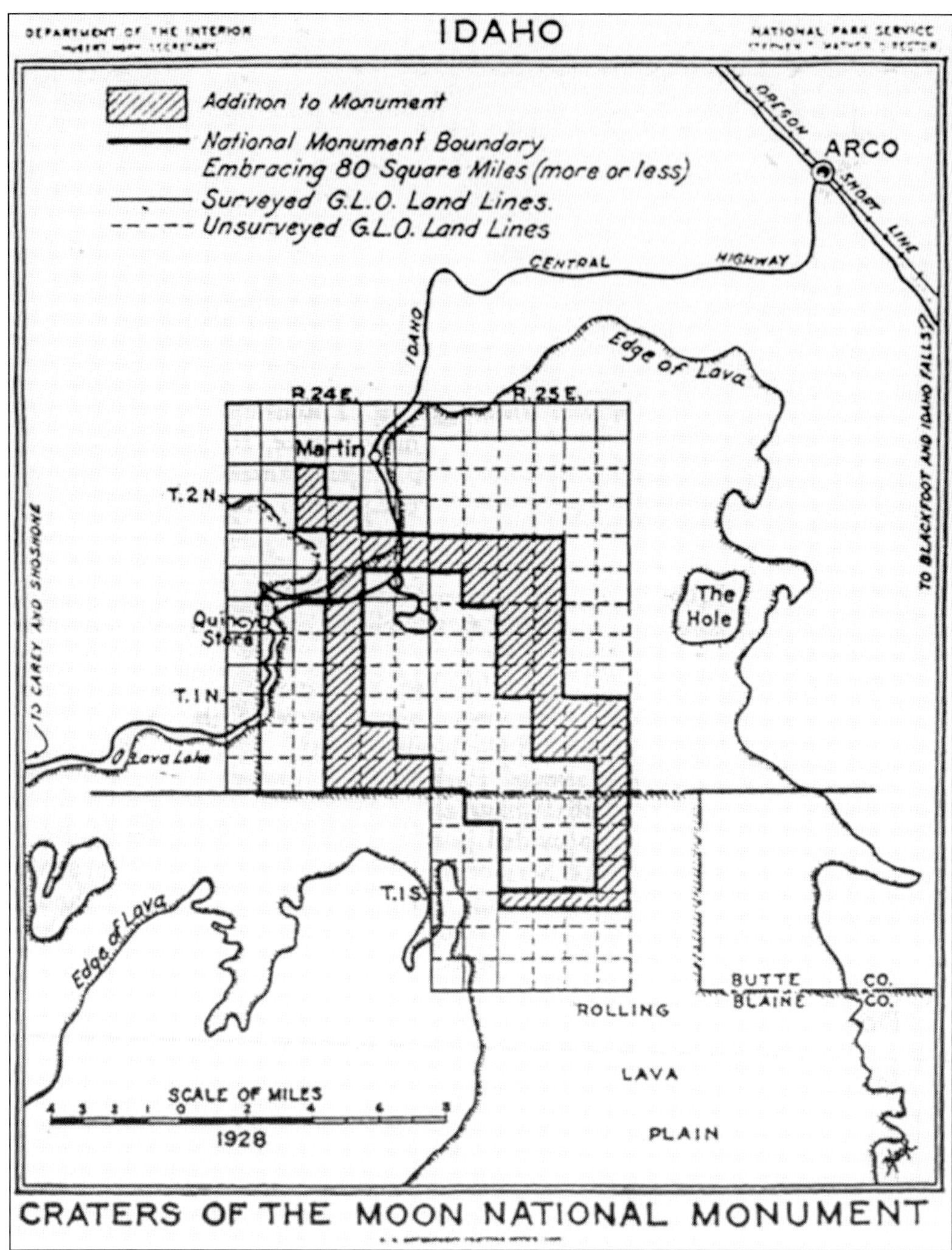

Immediately after the 1924 establishment of Craters of the Moon National Monument, it became clear that the original 39 square miles (101 square kilometers) would be impractical to manage without a reliable water supply. Geologist Harold Stearns was summoned to re-survey the area and recommend an expansion of the boundary. In 1928, President Coolidge accepted his recommendations and once again utilized the Antiquities Act to expand the boundary to include "certain springs for water supply and additional features of scientific interest." Through this second presidential proclamation, a portion of the area commonly referred to as the "North End" was added to the monument.

In 1962, President Kennedy added the 180-acre (73-hectare) Carey Kipuka and connecting lands to the southwest boundary of the monument. As described in the proclamation, "It was an island of vegetation completely surrounded by lava, that is scientifically valuable for ecological studies because it contains a mature, native sagebrush-grassland association which has been undisturbed by man or domestic livestock."

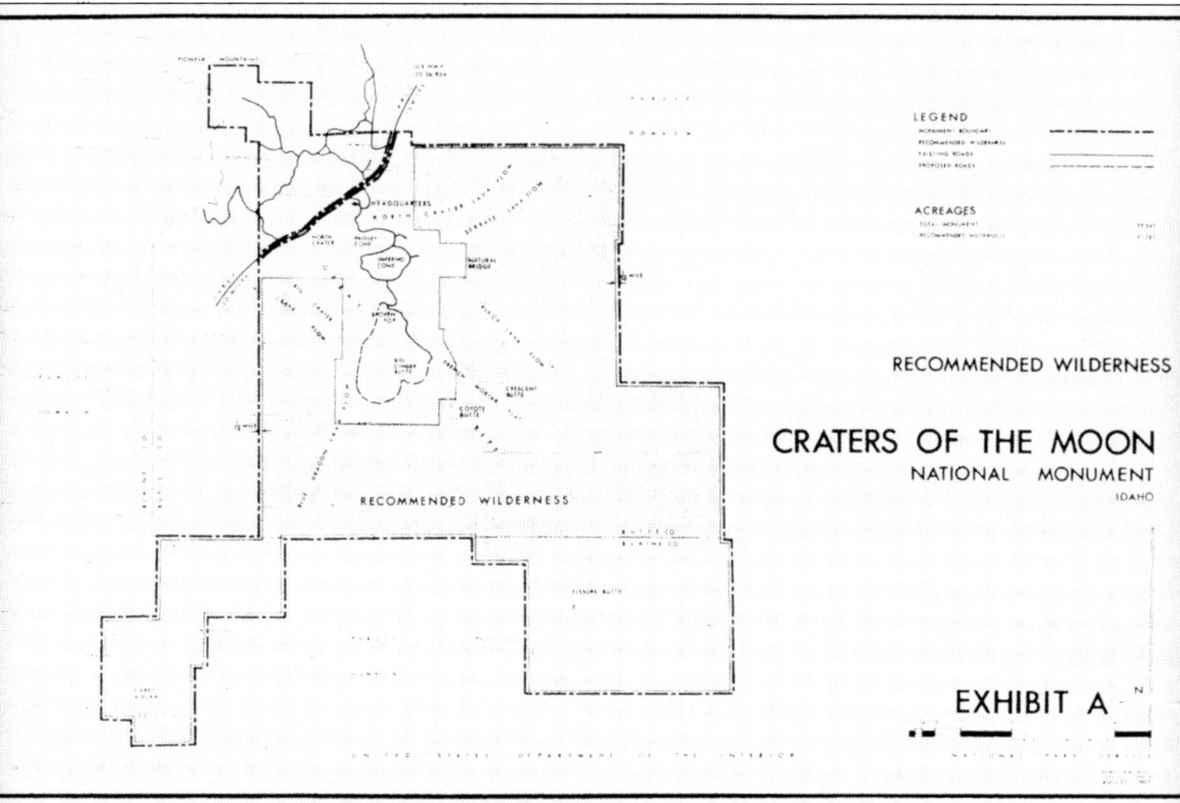

Congress designated the 43,243-acre (17,500-hectare) Craters of the Moon National Wilderness Area in 1970. This was the first wilderness designation in the National Park System. The final designation included an additional 2,243 acres surrounding Big Cinder Butte. As shown on the map, this area had been initially excluded for the possible development of an additional loop road around the cone. However, Supt. Paul Fritz insisted that future development should not be expanded beyond the existing roads and facilities.

A hiker takes in the view from Fissure Butte looking south toward Sheep Trail Butte, all within the Craters of the Moon Wilderness. The Wilderness Act of 1964 defined wilderness as "an area where the earth and its community of life are untrammeled by man, where man himself is a visitor who does not remain"—a perfect description for this vast area.

Federal Register / Vol. 65, No. 221 / Wednesday, November 15, 2000 / Presidential Documents 69225

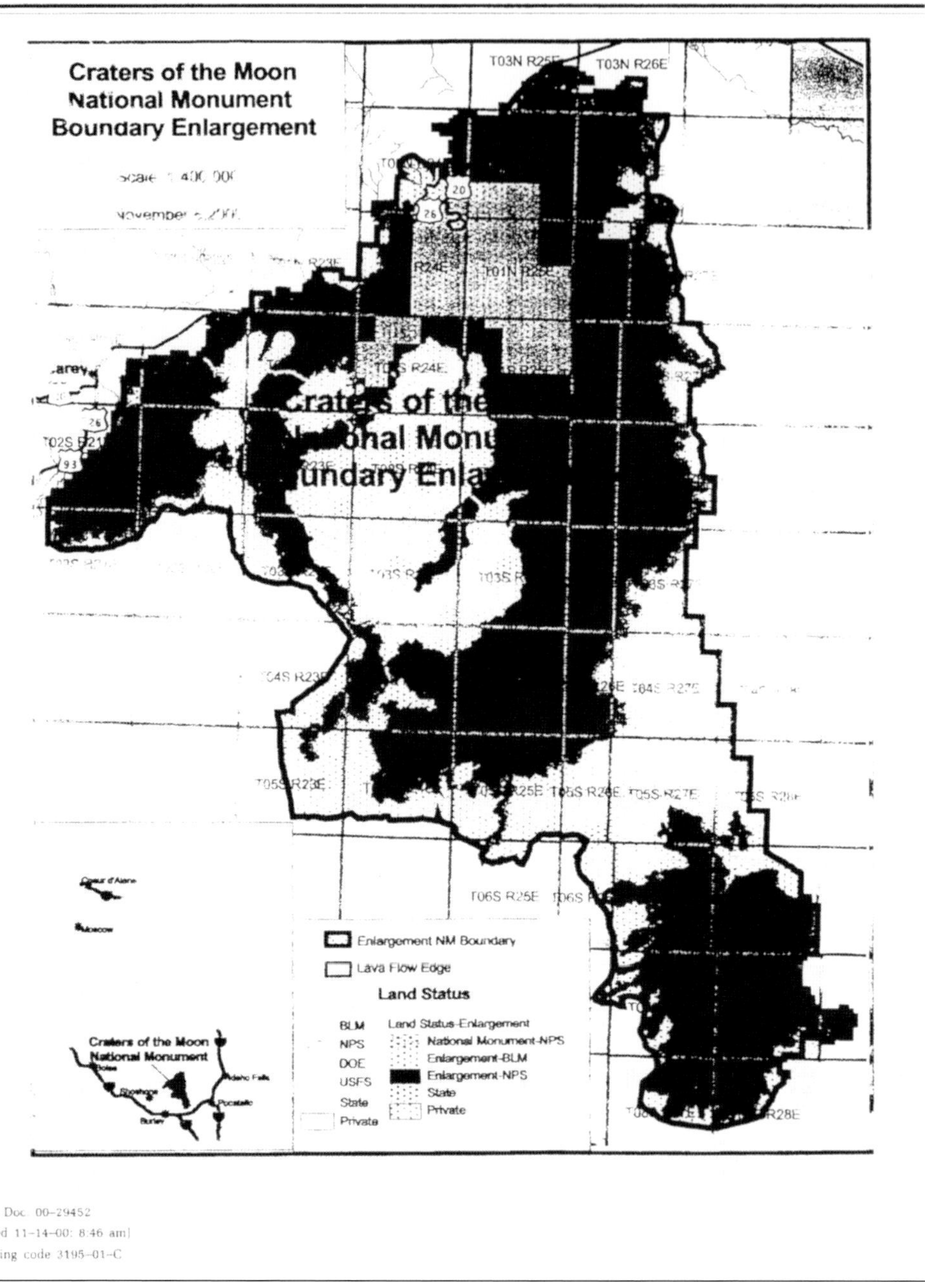

[FR Doc. 00–29452
Filed 11–14–00; 8:46 am]
Billing code 3195–01–C

In 2000, President Clinton used the Antiquities Act to expand the monument from roughly 54,000 acres (21,853 hectares) to approximately 753,000 acres (304,728 hectares) "to ensure protection of the Great Rift volcanic rift zone and its associated features." The proclamation also charged the NPS and the BLM to manage the expanded monument cooperatively, with each agency having management authority over separate portions.

Supt. Jim Morris (right), shown here with Secretary of the Interior Bruce Babbitt, worked closely with the administration and the BLM to facilitate this major expansion.

During Secretary Babbitt's visit, he received a tour from USGS geologist Mel Kuntz.

Babbitt (center, without hat) also met with other stakeholders, including local ranchers, to discuss the expansion. They were pleased to hear livestock grazing would continue on the lands where BLM would retain jurisdiction.

In 2002, Congress redesignated the 410,000 acres (165,921 hectares) recently added to the NPS monument as a national preserve. This act allowed hunting to continue on these former BLM lands.

Today's Craters of the Moon National Monument and Preserve consists of three adjoining administrative areas. Most visitors tour the 53,420-acre (21,618-hectare) monument, which is managed by the National Park Service. The dark, lava-covered area in the background of the photograph above is a portion of the 409,460-acre (165,703-hectare) NPS preserve. The sagebrush-grassland area in the foreground is part of the 274,800-acre (111,208-hectare) BLM monument. Hunting is permitted in the NPS preserve, and both hunting and grazing continue on lands managed by the BLM. (Both, courtesy of BLM.)

Approaching the visitor center from east or west on US Highway 20/26/93, visitors drive through many miles of lava terrain that are part of the NPS preserve leading to the NPS monument, where the visitor center and Loop Road are located.

The NPS preserve includes many outstanding features, such as Kings Bowl. This large crater is approximately 100 feet (30 meters) wide by 330 feet (100 meters) long. In this portion of the Great Rift, lava erupted explosively when it encountered groundwater about 2,000 years ago. The importance of this area was first recognized in 1968 when the Great Rift System was designated a national natural landmark.

Beginning in 1965, the Kings Bowl area was visited by thousands of people when the attraction known as Crystal Ice Cave opened to the public. The proprietors of this show cave, Jim and Peggy Pappadakis (pictured) provided access for paying customers via a tunnel they blasted through the rock. The tunnel led visitors 150 feet (46 meters) down into the Great Rift to views of fantastic columns of ice preserved within the interior of the cave. Unfortunately, increased air circulation from the tunnel also increased the melting of the icy formations, which led to unsafe conditions within the cave. These factors and the challenging economics of maintaining a business deep within the Idaho desert led them to cease operations in 1975. (Courtesy of BLM.)

This trailer served as the fee booth and operations center for Crystal Ice Cave during its heyday. Today, only the deep chasm of King's Bowl and interpretive signs remain to tell the story of this remote outpost in the preserve. (Courtesy of BLM.)

The BLM monument includes many miles of dirt roads winding through a vast area of sagebrush grasslands. Four-wheel drive and high-clearance vehicles are recommended for exploring this area. (Courtesy of BLM.)

Within the older lava formations of the BLM monument, there are several caves, such as the Bear Trap Cave system, shown here. The several segments of this lava tube extend almost eight miles (13 kilometers). (Courtesy of BLM.)

Most of the NPS preserve and some areas within the BLM monument have also been deemed suitable for wilderness designation and are therefore identified as wilderness study areas. In 1985, Pres. Ronald Reagan recommended to Congress that 322,450 acres (130,491 hectares) of the Great Rift Wilderness Study Area be formally designated as wilderness.

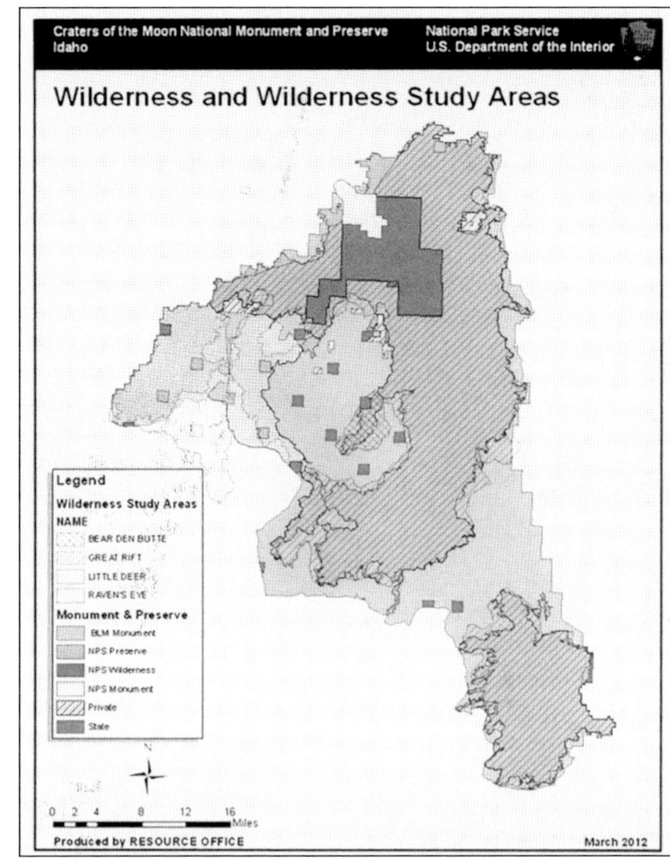

Policy requires that wilderness study areas be managed "so as not to impair the suitability of such areas for preservation as wilderness" until Congress determines otherwise. (Courtesy of Craig Wolfram.)

In April 2014, a team comprised of park staff and park neighbors hiked across Craters of the Moon National Monument and Preserve from south to north, following the Great Rift. The expedition was launched to investigate remote, rarely visited areas, and to celebrate the 50th anniversary of the Wilderness Act. The "Rifters," from left to right, are photographer Craig Wolfram, botanist Mike Mancuso, Supt. Dan Buckley, park ranger Ted Stout, Student Conservation Association intern Allie Konkowski, and Lava Lake Ranch owner Brian Bean. (Courtesy of Craig Wolfram.)

Members of the expedition encountered many hardships but also experienced the joy of discovery as they marked the location of previously unmapped caves and observed a wide variety of flora and fauna. They were dwarfed by the immense landscape, including the spectacular Sheep Trail Butte cinder cone, as seen here. (Courtesy of Craig Wolfram.)

In recognition of the outstanding night skies at Craters of the Moon and staff efforts to preserve them, the International Dark Sky Association designated the monument an international dark sky park in 2017. Later that summer, park staff announced the designation at a well-attended viewing of a total solar eclipse that passed through the area.

In recognition of NPS Modern architecture at Craters of the Moon, the Mission 66 Historic District was listed in the National Register of Historic Places in 2022. The district includes the headquarters area and the Loop Road. The facilities built or expanded during the Mission 66 period, 1957–1968, were critical to the development and public use of the monument. All of the structures, except for this original entrance sign, remain today and are largely unchanged.

Bibliography

Clark, D. *Idaho's Two-Gun Bob Limbert.* Arco, ID: Craters of the Moon Natural History Association, 2010.

Clark, D. and Craters of the Moon National Monument and Preserve. *Craters of the Moon: A Guide.* Arco, ID: Craters of the Moon Natural History Association, 2010.

Integrated Resource Management Applications (IRMA) Portal. irma.nps.gov.

Kuntz, M.A., D.E. Champion, E.C. Spiker, R.H. Lefebvre, and L.A. McBroome. "The Great Rift and the Evolution of the Craters of the Moon Lava Field." *Cenozoic Geology of Idaho.* Boise, ID: Idaho Bureau of Mines and Geology, 1982.

Limbert, R.W. "Among the 'Craters of the Moon.' " *National Geographic,* March 1924: 303–328.

Louter, D. *Administrative History: Craters of the Moon National Monument.* Seattle, WA: Cultural Resources Division, National Park Service, 1992.

———. *Historic Context Statements: Craters of the Moon National Monument.* Seattle, WA: Cultural Resources Division, National Park Service, 1995.

Owen, D.E. *Geology of Craters of the Moon.* Arco, ID: Craters of the Moon Natural History Association, 2014.

Ramacher, L. *Interpreting Cultural Resources at Craters of the Moon National Monument and Preserve.* Craters of the Moon National Monument and Preserve, National Park Service, 2011.

Stearns, H.T. *Geology of Craters of the Moon National Monument, Idaho.* Arco, ID: Craters of the Moon Natural History Association, 1963.

Stearns, N.D. "Exploring the Craters of the Moon, Idaho." *Bulletin of the Geographical Society of Philadelphia.* 26(4): 279–290, 1928.

Zink, Robert C. *Short History: Craters of the Moon National Monument.* Craters of the Moon National Monument archives, National Park Service, 1955.

About Craters of the Moon NHA

Craters of the Moon Natural History Association provides financial and staffing support to the National Park Service to educate and inspire the public to preserve and explore Craters of the Moon National Monument and Preserve.

The Craters of the Moon Natural History Association was established in 1959 by a concerned group of local citizens who wanted to make sure this special place in their backyard was protected. The organizational structure of the NHA consists of an executive director, a volunteer board, and bookstore clerks.

Purchases of books, like the one in your hands, from the NHA help to support the park through this vital partnership. The NHA sells publications at the Robert Limbert Visitor Center and online at www.cratersofthemoonnha.org.

Consistent with our mission to preserve history on a local level, this book was printed in South Carolina on American-made paper and manufactured entirely in the United States. Products carrying the accredited Forest Stewardship Council (FSC) label are printed on 100 percent FSC-certified paper.